OEE Guide to Smart Manufacturing

Dr. Jill O'Sullivan • Dr. Theresa Nick • Sandy Abraham

IMAE

A Division of DMMSI Associates, Inc.

OEE Guide to Smart Manufacturing

by Dr. Jill O'Sullivan, Dr. Theresa Nick, and Sandy Abraham

For Product Information, or Permissions as expressed above, Contact DMMSI at:
DMMSI Associates, Inc., 320 Carleton Avenue, Suite 6200, Central Islip, NY 11722
Website: www.instituteformae.com, Go to *Authors*

ISBN: 978-0-9912142-4-2
Library of Congress Control Number: 2016937038

Screenshots courtesy of: using MS Office Suite 2010, FORCAM, Inc.
Permissions granted by FORCAM® | Editorial by Dennis Goldensohn
Interior Layout/Cover design by Bernita McGoldrick/FORCAM

Disclaimer: All companies named in "Customer Story" segments are named fictitiously, and are used solely for purpose of explanation, not to promote any particular company.

IMAE, IMAE Publications, Institute of Manufacturing and Academic Excellence, And DMMSI are trademarks and Division of DMMSI Associates, Inc.

Table of Contents

Preface

For Whom is This Guide:

- If you are interested in addressing your shop floor management issues.

- If you're absolutely new to Overall Equipment Effectiveness (OEE).

- If you already have a continuous improvement process, a Lean initiative in your company but need answers to machine efficiencies for company specific issues.

- If you are considering a new shop floor management tool to measure Overall Equipment Effectiveness (OEE) at your company.

- If your company is improving their Total Productive Maintenance Program (TPM), this OEE Guide would be good for understanding performance of the equipment on the shop floor, then **this Guide is for you.**

Conventions Used In This Guide

OEE Guide Featured Symbols

Throughout this Guide, you will find symbols that will help you better understand the material, with the ability to pinpoint ideas and thoughts to discuss.

 This symbol indicates an OPPORTUNITY to challenge current thinking.

 The POLICY symbol indicates the need for review of a Policy.

 The IDEAS symbol prompts to think of new ideas and customer focused insight.

 The SCALE symbol asks the reader to consider the importance of the statement.

 The STOP Sign symbol indicates an important point to be understood.

 BE AWARE - and contemplate warning.

 TOOLBOX for specific tools.

 USING EXAMPLES: Real world examples to better understand existing issues and challenges to find successful solutions.

DMMSI IMAE Onl ine
www.instituteformae.com

DMMSI/IMAE provides publications that deliver expert content in text and e-text form from the world's leading authors in Shop Floor Management (SFM) Overall Equipment Effectiveness (OEE) technology.

About the OEE Guide

Foreword

OEE Guide to Smart Manufacturing

How manufacturing companies use their production resources to bring their products to market is what separates and differentiates them from their competition. Having the highest level of responsiveness on the shop floor in order to meet the customers demand while being lean and continuously improving, is key in this very competitive environment. There is a great need for target oriented performance measurement and benchmarking amongst the fierce competition in every industry in manufacturing today.

Shop Floor Management (SFM) is one of the most important ongoing activities performed in every manufacturing organization. As stated by Kiyoshi Suzaki, author of The New Shop Floor Management, "the shop floor is revered as the place where people ultimately add value to their society and strengthen its foundation." Good shop floor management, SFM, tools protect a company's lean implementation initiative and promote culture transformation.

This OEE Guide will break this down into simple and manageable steps using an SFM tool for better efficiencies. The Guide will evaluate changes in BIG DATA, issues in manufacturing including; Shop Floor Management; Availability, Utilization, Quality, Performance, Measurements, Metrics, Downtime, Order Management, Scheduling, Machine Data, Synchronization and ERP.

This Guide addresses the issues of OEE, discussing both reliability and maintainability and exploring the overall effectiveness of machines as well as the human element in a manufacturing environment.

Modern Shop Floor Management supports all roles in production from the machinist, the maintenance technician, factory operators and managers. This establishes a closed loop communication that delivers the right information to the right person at the right time. Regular Continuous Improvement Process, CIP, meetings provide the analysis of objectives and targets safe use with real time factory information. Having a guide to assist in this very important function will allow for increased and continuous improvements.

This guide uses key points of Overall Equipment Efficiency, OEE, ideal for operators, mechanics and team leaders/management for improving effectiveness and overall efficiencies.

About the Authors

Dr. Jill A. O'Sullivan is an Assistant Professor at Farmingdale State College in NY, where she teaches ERP, MIS, Operations and Supply Chain Management. As the first past president of APICS NYC/LI she serves as the Faculty/Student Advisor of the Student Chapter at Farmingdale State College. She is on the Advisory Editorial Board of the Journal of Systemic, Cybernetics, and Informatics (JSCI). Dr. O'Sullivan is an original member of the Manufacturing Executive Leadership Council.

Founder/President of JJKT Consulting Services and previously founded JJK Sales, a Manufacturers Sales company representing the northeastern regions specializing in custom manufacturing, assembly and electronic component products for Industrial/OEM, Distributor/Wholesalers, Value Added Reseller, Military/Aerospace, Government and Medical companies. Formerly a Purchasing and Materials Manager on LI, Mrs. O'Sullivan has acquired more than 24 years of experience in Operations and IT in military, aerospace, government and commercial companies. Jill and her husband, John, have two sons, Kyle, Tom and Oliver, the family dog.

Dr. Theresa L. Nick is Engagement Manager at FORCAM, Inc., a Manufacturing Execution Process (MES) software provider, responsible for deploying the software at their customers' facilities. Additionally she trains the company's customers in engaging with the software and using it to improve productivity on their shop floors via Continuous Improvement Processes.

Her background is in the area of Information Technology where she published numerous articles in a diverse range of journals and presented her research at different conferences worldwide. Theresa's new focus is on the benefits that come with real-time information about shop floors in any kind of industry.

Born and raised in Germany she is currently living in the US together with her husband, Sebastian.

As an international marketing specialist leveraging global corporate and institutional experience and an upbringing spanning multiple countries, Sandy Abraham has built a strong ability to collaborate, facilitate and innovate every aspect of business.

Sandy Abraham is a true team player who leads all global marketing activities and brand management of FORCAM. Passion for technology and innovation drive her enthusiasm to lead marketing teams in the Asia-Pacific, Europe and the Americas.

~Acknowledgements~

We would like to thank FORCAM, DMMSI/ IMAE and JJKT Consulting Services for their insight and support of this guide. Our sincere thanks!

Chapter 1
Shop Floor Management

Chapter 1 emphasizes how manufacturing companies use their production resources to successfully bring their products to market which is what separates them from their competition. Having the highest level of responsiveness on the shop floor to meet the customers' demand, while being lean and continuously improving, is key. Effective production requires precise knowledge of which machine is providing what at any given time. This provides an overview of the consumption of energy and material allowing for the company to keep track of how much time there is in each step in an effort to reduce production costs. It is important to have the tools that ensure mutual cooperation of everyone in the organization associated with all business processes. This enables a company to meet and hopefully exceed the needs and expectations of their customers.

 Shop Floor Management (SFM) is one of the most important ongoing activities performed in every organization.

The Core Elements of Overall Equipment Effectiveness

Overall Equipment Effectiveness (OEE) is a crucial measure in Total Productive Maintenance (TPM) that reports on how well equipment is running. TPM seeks to engage all levels and functions in an organization to maximize the overall effectiveness of production equipment.

OEE Factors these Three Elements:

Availability = Run Time / Total Time
- The time the machine is actually producing / the time the machine is adding value to the product.

Performance = Planned Cycle Time/Actual Cycle Time
- The quantity of products the machine is turning out in a certain amount compared to what was planned.

Quality = Good Count/Total Count
- The quantity of good output – into a single combined score.

The Formula used for OEE is: OEE = Availability x Performance x Quality

Some manufacturers believe the formula for their operation is OEE=Value-Add x Quality Factor

- **Availability** = Run Time / Total Time
- **Performance** = Planned Cycle Time / Actual Cycle Time *(based on a standard)*
- **Quality** = Good Count / Total Count

For some industries, World Class OEE is 85%.

Depending on a customer's specific industry it might be highly unlikely to reach an OEE of 85%, but the willingness to increase productivity should be the goal of any facility. Even an increase in OEE by only some percentage can mean meeting demand and increasing profits.

The Following are Primary Core Elements of OEE:
Systems and Techniques

Identification of the relevant tools and techniques pertinent to each different stage of manufacturing is required. This includes the area/project and the conditions in which the tools should be used to achieve successful application.

The tools used should be identified as familiar to employees and be classified as core or optional depending on their nature and impact each has on the working environment. Some of these tool include; Gemba, value stream mapping, cause and effect diagram, Pareto analysis and others. Work should be viewed as activities, connections and flows.

In Assessing Business excellence (Tanner, 2004), they outline that self-assessment results when set against an excellence framework provide organizations with insights to what their strengths and weaknesses are.

Correctly integrated information, allows manufacturers to significantly increase productivity and gain dramatic increases in production throughput while reducing MUDA, a Japanese term for waste. (See Glossary)

Processes

Analysis of processes should be an integral part of any manufacturing company. There should be a focus on processes rather than the functions, and be part of the Kaizen initiative for continuous improvement, CIP. (See Glossary) In Japan they have a saying, 'Look after the process and the product looks after itself!' (Hutchins, 1990) This should be an integral part to the development of best practices ultimately encouraging a Kaizen environment. Practicing personal Lean is the way to go.

The ultimate goal is to demonstrate the:

- Value of real-time
- Accurate data reflecting production activity
- Planned production schedules
- Quality checks
- Machine breakdowns
- Inventory levels through improved processes using the OEE tool.

Knowing every second what state an asset is currently in can be extremely beneficial. Seeing for example that a machine is in dry-run for 6 seconds every minute in a high volume production environment can lead to increasing productivity by 10% when the machine speeds are adjusted to be constantly running and not waiting for material.

Management

As the corner stone of a successful OEE program, Management holds the seat of most influence to the employees working for the company. They ensure that they get results through their employees. A company's management style is important when trying to get the most out of their employees, whether it is through an authoritarian approach or a participation styled approach. Management must be OEE driven for success to occur. Their commitment and personal involvement in OEE helps in creating and deploying well defined systems, methods/standards and performance measures for achieving their OEE goals. They encourage participation by all employees and a change of culture in the company. Beliefs, behaviors and performance have to be addressed. People need to change the way they see things in order to change.

People

People are considered one of the most important resources of a company. They are those productive resources that use their talents to assist the company to produce their parts and or services. Yet, they could be in the wrong spot doing the wrong job. Remember, as Jim Collin says in his "Good To Great" book, "Whose driving the bus?" Is it the trained bus driver or someone who does not know how to handle this very large powerful vehicle? Can we assess that the employees are in the right position for the maximum effectiveness of the manufacturing process? Additionally, change in a process, of a system, technology or of a way an employees does their job affects everything for them. Having the employees motivated to contribute will provide more success; gone are the days or brute force. Overall Equipment Effectiveness (OEE) will happen with employee engagement. Shop floor operators monitor and track the data and provide important inside information. Companies struggle to encourage contribution of individual employee input in a way that they can take ownership of the process. Those involved need to be problem solvers to eliminate the issues they witness each day on the job.

When implementing change, companies cannot expect that all employees will accept this initiative because management says it should be so. It is imperative that management keep employees abreast at all times when decisions are being made, which should encourage participation and help ease transition tension. Training is one of the most important motivators besides having the employees involved in the decision. Employees will do much better if they are educated in the subject area prior to training. Allowing for employees to better understand the cross functional relationships between their departments and those that feed to or rely on that department is crucial. The internal 'upstream' and 'downstream' activities are dependent within a corporation. No longer do we have the luxury to sit in silos when our competition is moving away from the older management scenarios.

 Without proper training, no matter what the motivation, the OEE tool will fail. As Carol Ptak says in her book "ERP, Tools, Techniques and Applications for Integrating the Supply Chain", "Labor was the driving force for productivity. Empowering Employees was needed for providing the agility that was required to compete in the market". Training given to the right people has proven to minimize the misuse of the tools and techniques; Ownership, Empowerment and Self –Directed Teams.

Employees are encouraged to take more responsibility, communicate more effectively, act creatively, and be innovative through on-going education and training. Some tools for improvement are very expensive; therefore, it is downright foolish not to train the employees to use the tool correctly. Education is primary and training secondary. As a result, employees will take on ownership, and will feel empowered through self-directed teams to mitigate problems that occur on a daily basis on the shop floor. This will provide the maximum opportunity for the efficiency desired. After the investment of education and training there should be a return on investment (ROI) that is measured.

Teamwork

Teamwork needs to be emphasized in the OEE initiative. This requires all employees to work individually and as a team whether in one department of a company or interdepartmentally. In a continuous improvement environment, teams could develop the culture through building collective responsibility to develop a sense of ownership. Teams provide additional communication channels between individuals, supervisors, management, customers and suppliers. They develop problem solving skills and facilitate awareness of quality improvement potential, leading to behavioral and attitude change. (Barrie G. Dale, 2007)

Teamwork for successful SFM is dependent on the successful understanding of roles, relationships and responsibilities. Utilizing new technology for the increased improvements available by using an SFM tool all hinge on, skills acquired and interrelationships between management and workers.

Figure 1.1 Shop Floor Visual © 2015 FORCAM

Management

○ Operating state development
○ Availability of machines
○ Dashboard and availability report

Manufacturing Engineering

○ Operating state development
○ Availability of machines
○ Dashboard and availability report
○ Hit list of stoppages

Operators

○ Runtime protocol
○ Runtime monitoring
○ Operating state development
○ Availability of machines
○ Dashboards and availability report

Maintenance

○ Hit list of common stoppages

Culture

A culture of Lean Manufacturing has to set the standard for discipline, efficiency and effectiveness. Therefore, when we look at the culture for OEE, we always look at Lean with the same promise of the standard for discipline, efficiency and effectiveness. OEE does not require a radical change in the company's culture if they are already striving towards a leaner organization to produce their parts and service their customers.

 Continuous Improvement Process (CIP) is a fundamental factor in leadership; includes philosophy, style, and behavior which need to be aligned with human resource systems.

This may include: job design, (job description/ specifications) selection processes, compensation rewards, performance appraisal, policy, procedures and training, and development.

Success Factors

To obtain success with any specific tool that a company purchases will depend on crucial elements, as described below:

- First, the tool must fit the needs of the organization, and its intended plan of implementation. Therefore, a detailed analysis should be completed prior to confirming the organizations intentions with hopes of meeting the expectations and goals.

- Second, the entire company from the top down has to be on board as to its purpose, the need for the tool and the expected results.

- Third, the company must train employees to use the tool effectively for improvements in their functional area. Lack of education and training is a major reason behind failure! Poor initial planning, not truly recognizing the resources that will be required, inaccuracies in the projected budget required and the misallocation of time will lead to failure.

 Success does not just come because a company purchased a new fancy tool. This takes a mind-set of continuous improvement (CI) and a collaboration effort by the entire company especially those in the functional areas that the tool will be used in. Leaders have to be committed, from their perspective, that this is important for the company to resolve issues and to be more efficient.

OBJECTIVES IN OEE IN SFM FOR INITIAL SUCCESS

- Create a steering team among members of the workgroups to help strive toward consistency.
- Create standard work used throughout the plant.
- Establish appropriate critical metrics, (Key Performance Indicators – KPIs) specific and relevant to each area.
- Decide necessary supporting documentation for each area. SOP - Processes work instructions.
- Provide training in roles/responsibilities for all key players; Team members, Team Leaders, Supervisors, Managers, Senior Managers.
- Usage of Visualizations and Reports.
- Establish a pattern, consistency, and quality of routine meetings.
- Daily stand up meetings (huddles) between team members and leader.
- Weekly Report Out (CIP) meetings between the team leader and management.
- Cultivate and reinforce the idea / thought process of:
  - Daily Schedule Attainment / Goal setting & achieving.
  - Daily striving / continuous improvement.
  - Celebrate successes and reflect on the outcomes.

Benefits of SFM

This is the best of Visual Manufacturing systems:
- Visual systems for Team members, Team leaders, Group leaders, and Managers.
- Transparent data collection (nothing is hidden).
- Access to the data anywhere anytime.
- Determines the key data to collect to better manage.
- Establishes escalation process.
- Establishes standard work practices (a pattern of consistency and quality for routine meetings).
- Increased communication efficiency.
- Clarity of company oriented organization.
- Increased involvement of everybody.
- Better problem solving capabilities.
- Greater value for the end consumer.

Having a Manufacturing Execution System - MES deployed unfortunately does not mean that OEE will increase automatically. Displaying the data captured on the shop floor where everybody can see it, this is an important first step. It raises questions from operators as well as supervisors. Once people are talking about potential problems, ideas for improvement start coming up at the same time. However, analyzing the data before starting with any new procedures should always be the first step. The collected data provides so much information to determine a root cause of many problems.

Seeing that a machine is on a high frequency in 'Program stopped' does not automatically mean that the operator is doing something wrong, but maybe his experience tells him that the given program does not take into account some specialties of the machine. A collaboration between experienced operators and NC programmers can lead to improved programs that run without operator interference, giving the operator the chance to attend more machines at the same time and leading to an increased availability (and most likely also performance and quality) for the machine. This leads into plant floor layout considerations whether two machines can be run by one operator.

PRINCIPLES (LEADERSHIP, TOOLS, MANAGEMENT STYLE)

Traditional Vs. Advanced Shop Floor

In the past, the main goal of a shop floor was to produce a good product. There was a given process how this was done, all information needed was printed. When a part was not produced in the time scheduled, there was no information right away why the goal was not achieved. Decisions for changes were made by the management.

Over time it was realized that it is essential to make changes where the actions takes place (Japanese "Gemba"), i.e. the shop floor became the center of attention. The need for real-time data about ongoing processes is inevitable in today's production flow which is driven by the lean principles of on demand production and just in time delivery.

Based on those needs, other factors that are hindering efficient production became obvious: data needed for the process was not always available when needed, i.e. what order to produce at what time with what material and tools.

These observations lead to the development of Manufacturing Execution Systems (MES) that are able to monitor production assets (machines), provide information about orders and also include additional functionality like finite scheduling capacity and assisting paperless production via digital availability of bills of materials and production files, e.g. drawings.

 In many countries initiatives were formed to accelerate those ideas and create standards. Examples are the "Industrial Internet Consortium®" in the US and "The New High-Tech Strategy" from the German federal government just to name a few.

CONTINUOUS IMPROVEMENT PROCESS (CIP)

Within a company, at all levels in all departments, a mindset of continually improving should be primary, including all of the tools and techniques that should be clear to the users. Recognized attainable goals and standards should be set with traceability through feedback. This will provide an opportunity to analyze what happened and how it could be improved, the retrospective opportunity to learn and change. The continuous improvement process or Kaizen mind-set will be improved through daily experimental actions. A company should consider a Help Desk/ Ticketing system to capture all data of successful experiences and their origins.

Figure 1.2 PDSA

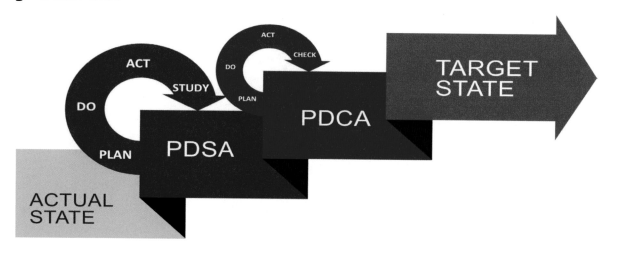

A good way to achieve continuous improvement is through always having so-called Plan-Do-Study-Act / Plan-Do-Check Act cycles running. This is an approach to look at the current state, analyze it with the data provided by your MES and come up with ideas for improvement. The importance though is to not stop after implemented for example a new process, but to follow up and check if it is effective, i.e. achieving an improvement. The final step then is to act accordingly which either means you standardize it and roll it out throughout the complete facility if it is improving your productivity or if it is not, you need to revise and try to come up with a new idea.

 A PDSA cycle can be done on various different topics, some can be even ongoing in varying stages in parallel. They can be achieving minor or major improvements. However, the importance lies is running through the complete cycle and not stop after implementing it without checking its effectiveness.

CIP TEAM AND CLOSED LOOP COMMUNICATION

As an introduction to OEE and the benefit of its use on company machinery and with the company's ERP system, we divulge areas of concentration in this section for better comprehension and efficiency in your company.

Areas in a company that embrace OEE and elected to learn from the analysis, look for ways to improve their business based on losses. This then demonstrates how to improve the process and more importantly, how to gain a reduction in the product's cost.

The important thing about OEE is to use the measure to understand your business and equipment for improvement. While it is possible to determine the cost implications of each of the OEE factors, OEE provides a measure to see how effectively an asset is utilized to produce a quality part where time (not cost) is the baseline!

The early Toyota Production System (TPS) focused on "eliminating waste to reduce cost." OEE was initially developed to identify the "major losses" in equipment performance and reliability. "All we are doing is looking at the time line, from the moment the customer gives us an order to the point when we collect the cash. We are reducing the time line by reducing the non-value adding wastes." - Taiichi Ohno - See more at: http://matthrivnak.com/2008/05/05/taiichi-ohno-quotes/#sthash.hinAZZnE.dpuf.

TPM then became a company-wide approach to eliminating the major equipment losses. OEE addressed whether the equipment was doing the right things.

Some Major Losses

Availability Losses

- Planned Shutdown Losses
    - No production scheduled (i.e. no shifts or breaks)
    - Planned maintenance

- Downtime Losses
    - Breakdowns & failures
    - Changeover (product, size) and setup
    - Tooling or part changes
    - Startup (first piece) or adjustment

- Organizational Losses
    - Meetings & trainings
    - Waiting for material, tools, IT, engineers, quality control etc.
    - Cleaning
    - Transport

- Performance Efficiency Losses
    - Minor stops (jams, circuit breaker trips, etc.)
    - Reduced speed, cycle time, or capacity (runtime)
    - Non-standard routing (operations without standard set up and run time)

- Quality Losses
    - Defects/rework
    - Scrap
    - Yield/transition (from changeover, startup/adjustment)

After deploying a MES system and collecting data for a few weeks, the first thing to do should be to calculate a baseline containing average times for production as well as losses. Based on this information, a decision can be made identifying what the major losses are on a shop floor. While some companies have to face breakdown reduction, others need to update and improve their NC Programs while still others need to think about improving their material and/or tool handling. It might become obvious that handling two machines at the same time is not possible for one operator because the cycle times are too short leading to waiting times on both of the machines because the operator is always busy at the other machine.

After collecting the data as prescribed above, for several weeks, you could then perform a Pareto analysis to see which one of the items are causing the most inefficiencies in the machine shop. If 20% are causing 80% of the problems, this is what is called, low hanging fruit that can be picked for immediate cycle reductions. By experimenting and employing solutions, you can validate that the solution has eliminated the cause for reduced cycle time. This solution will become standard work in the machine shop. (Plan, Do, Study, Act/ Plan, Do, Check, Act).

SUMMARY

Even though you have this system, if you have people who do not have the education or training to comprehend the true meaning of the data, the tool will not meet your expectations. OEE data collection, analysis, reporting, and trending provide the fundamental underlying basis for improving equipment effectiveness by eliminating the major equipment-related losses. Shop floor machine data used in root cause analysis, with the identification of the specific root causes leads to finding solutions on a daily basis. OEE data very quickly leads to root-cause identification and their elimination.

OEE data then answers the question, "Did we eliminate the root cause of poor equipment performance?

Customer Success Story—Richards Industries

by Modern Machine Shop, Mark Albert, Editor-in-Chief

April 1, 2016

Getting Started with Machine Monitoring

This shop's successful entry into machine monitoring reveals important points about what to do and what to expect.

Machine monitoring provides a window into the performance of a machine tool. Here, VP of Operations Bill Metz leads a discussion about improvements in the use of a pallet changer on an Okuma MA500 HMC. With him are Bob Linville, manufacturing manager, Joe Dababneh, CNC operator/programmer, and Bob Luthy, continuous improvement manager. Data on the screen help pinpoint uptime gained by rapidly getting palletized work in and out of the machine.

Have you noticed that some of the skinniest, fittest people you know are now wearing watch-like devices to track their activities, such as the number of steps they take, the calories they burn and hours of sleep they get? Having this data at hand (or on the wrist, to be exact) increases their awareness of habits that promote a healthy lifestyle. Because they waste less time sitting around or not getting restful sleep, they tend to have leaner, stronger physiques—at least that is their goal.

In a real sense, manufacturing companies are thinking and moving along the same lines. Companies that are already practicing lean manufacturing techniques are turning to machine-monitoring systems to help them reinforce the habit of continually trimming down on activities that don't add value for the customer.

For example, Richards Industries, a Cincinnati, Ohio, company that manufactures industrial valves, has been practicing lean manufacturing for many years. Now, it has installed a machine-monitoring system that is enabling shopfloor personnel to track activities and record the performance of its machine tools. Like readings from a Fitbit or Jawbone, the data gathered and analyzed by this system is making the company more aware of how well machine time and manpower count toward productivity.

Although Richards Industries is still in the early stages of implementing this machine-monitoring system, the results are encouraging. They show that the company is moving to a higher level of lean operation. Machine uptime has increased significantly; a pallet-changing system is getting better usage; CNC programs are running more efficiently; and setup procedures have been further streamlined.

The valves and related products manufactured by Richards Industries are specially engineered for critical applications in a variety of industries, including chemical, petroleum, power generation, biotech and pharmaceutical. The company offers six distinct product lines that cover many types of specialty valves and related accessories, such as control valves, pressure regulators, instrument valves, manifolds, ball valves and steam traps.

Most of these products, which are marketed globally, can be ordered as standard items, but the company offers customized variations that are engineered for critical or unusual applications.

Richards Industries has manufacturing facilities in Cincinnati (its headquarters), but also sources some items from China, India, Taiwan and other countries. All customized and engineered products are produced in Cincinnati. In the machining area, occupied by about 30 new and older CNC mills and lathes (and a few "legacy workhorses"), batch sizes are small. Five to eight pieces comprise an average batch. A few one-offs and an occasional run of 100 or more pieces can also occur.

Keeping machines running productively and workflow moving smoothly is essential. Setup time is a major concern. Years ago, the company made setup reduction a primary focus of its long-standing commitment to lean manufacturing. "True lean manufacturing requires a constant focus on improvement in all areas that are preventing us from being productive, such as any activities that keep machines from cutting metal," Bill Metz, vice president of operations, says.

About three years ago, Mr. Metz and other managers at Richards Industries began to eye machine-monitoring systems as a better way to keep tabs on what the machines were doing, or not doing. They saw the promise of connecting machines to a network that could automatically collect and feed machine data to applications for analysis and reporting.

"We wouldn't have to suspect there was a problem before taking a closer look, we could spot the problem and take corrective action more quickly," Bob Linville, manufacturing manager, explains. However, the management team at Richards Industries quickly began looking beyond machine monitoring as the main value of a connected, networked shop environment. Two other broad and long-standing goals were included as essential parts of this initiative. First, managers wanted to automate the order-data management process so there was less reliance on operator input to capture timesheet entries, inspection results and details about machine utilization. Second, managers wanted paperless, networked distribution of part drawings, workpiece and material sheets, tooling lists, inspection sheets, and setup instructions, along with streamlined downloading of CNC part programs.

At the beginning of 2015, Richards Industries officially launched its machine-monitoring initiative and by June had a pilot program of 10 machines connected to a monitoring system for machine-data collection, data visualization on the network and automatic alerting. Another 18 machines are currently in the process of being connected. When completed, this machine-monitoring system will constitute Phase 1 of the long-range plan. Phase 2 will be the implementation of order-data management. Phase 3 will be product-data management and enhanced DNC. With the 10-machine pilot project in operation for about 10 months now, Mr. Linville confidently says the effort has been a success—with tangible, measurable results that "prove an improvement." He says that overall equipment effectiveness (OEE), which factors in productivity, performance and quality of output, is at least 20 percent higher than before machine monitoring was in place.

With the 10-machine pilot project in operation for about 10 months now, Mr. Linville confidently says the effort has been a success—with tangible, measurable results that "prove an improvement." He says that overall equipment effectiveness (OEE), which factors in productivity, performance and quality of output, is at least 20 percent higher than before machine monitoring was in place.

To encourage and help other shops move toward machine monitoring—and more—Mr. Metz and his colleagues have these recommendations and advice, which fall into five topic areas:

1. Have vision and commitment at the top.
2. Use a pilot program to set the course.
3. Reinforce shop culture with good communication.
4. Rally the team around an enthusiastic leader.
5. Get good data, and act on it appropriately.

Commitment at the Top

Richards Industries has been manufacturing specialty valves at its Cincinnati location since 1947. It has a history of company ownership that looked for ways to improve its manufacturing operations, its service to customers and the quality of life for its employees in the workplace. Its embrace of lean manufacturing and setup reduction almost 15 years ago is a prime example of this forward thinking. Concern for workforce relations and well-being is attested by the fact that the company has been voted one of the best places to work in the Cincinnati area, recently receiving top scores in six of the past seven years.

"One reason lean manufacturing made a difference here was the commitment of Gilbert Richards, then-owner of the company, to implementing it in earnest," Mr. Metz says. The company is now owned by a number of its top managers, including Mr. Metz. They have not forgotten this lesson.

Today's ownership group is solidly behind the move to machine monitoring. In fact, the impetus to get started can be traced to discussions among this group when machine monitoring first took on its current momentum with the introduction of shopfloor interoperability standards such as MTConnect about eight years ago.

Mr. Metz says that this essential "top management buy-in" has to be more than words. Positive actions are also necessary.

For example, shop owners must be committed to making the monetary investment. This means fully analyzing, reviewing, justifying and funding the cost of a machine-monitoring system. The budget should cover acquiring the system, installing the infrastructure, devoting workforce resources and training users thoroughly. The return on this investment has to be calculated, without losing sight of overall value.

"We figured that the most conservative estimates of productivity gains from machine monitoring would be about 5 percent. We used this figure to justify the cost of the machine-monitoring system, and it has paid off," Mr. Metz says. That said, he notes that the return on a monitoring system is not derived from the system itself, but from the actions that help operators and machines become more productive.

Likewise, it is important for top managers, supervisors and leadmen to be involved in all aspects of planning and implementing a machine-monitoring system. "This is best done with a team approach," Mr. Linville suggests. For example, early in the process, Richards Industries created a committee that included representatives from management, shop supervisors, leadmen, IT personnel and manufacturing engineers. This key group was heavily involved in the choice of the software provider with which to partner, and it will remain heavily involved through implementation of Phases 2 and 3 of the plan.

Plan Big, Start Small

Perhaps the most important decision Richards Industries made was to think beyond machine monitoring. By determining what other benefits could be derived from the monitoring network, the company realized it could go farther and get there faster with its Phase 2 and Phase 3 concepts in mind at the beginning. For this reason, Richards Industries selected Forcam, a software firm specializing in systems for discreet parts manufacturing that take a unified approach to achieving improvements across all operations, as its supplier for the machine-monitoring system. This decision was made largely on the basis that this supplier offered additional software products and capabilities for providing order-data management and product-data management/DNC.

Based in Germany, where Forcam has had considerable success as a systems integrator for larger companies in the automotive industry there, the software company chose Cincinnati as its U.S. base with the intent to focus on systems for small- and medium-sized manufacturing companies in this market. "Once we had Forcam on board, we could get their system engineers working with us to set objectives, create a timeline and establish criteria for how data was to be collected and used," Mr. Linville explains.

The 10-machine pilot program represents Richards Industries' initial installation of the Forcam Force Shop Management suite of software tools for machine-data collection, visualization (displays of pie charts, performance bars and trend lines), report generation and issuance of alerts. The first step was conducting a connectivity audit of all machines in the shop, then selecting the 10 with the heaviest usage or most critical operations that could become bottlenecks. The machines selected also represented the variety of adapters or interfaces needed for all of the company's machines. Getting these connections up and running during the pilot program would greatly facilitate shop-wide network implementation, which commenced in January 2016.

Half of the first 10 CNC machines were equipped with CNCs that were MTConnect-compliant (MTConnect is the machine tool interoperability standard that translates machine tool data into a common, Internet-based language). Forcam and Richards Industries worked together to filter the stream of MTConnect-formatted data so that only selected items for relevant reports were captured. One of the 10 machines required an IBH Link interface for formatting and tagging the data generated by its Siemens control system. Four other legacy machines required installing a Wago data-collection device that is connected to the input/output (I/O) board inside the control unit. These older CNC machines ran software with limited or nonexistent data reporting, necessitating that machine status and other operational events to be collected from the I/O signals.

Mr. Metz also emphasizes that it is critical to have the IT department involved from the very beginning of the process. "The support of Jeff Howard and Beth Mogg of our IT department has been outstanding. Their efforts have been critical to our success, and we will rely are on their involvement until the program is completed," he says. For its part, the IT department was responsible for having electrical power lines (for backup purposes) and network cables connected to all 10 machines. Network security measures such as "firewalls," password protection and encryption were also under the IT department's jurisdiction.

Communicate, Communicate, Communicate

Aware that a machine-monitoring system needed acceptance and commitment on the part of the shopfloor workforce, managers at Richards Industries made good communication a top priority during the planning and installation of the system.

The plans to implement a system were communicated in quarterly update meetings for more than a year, before installation began. Once the decision was made to go ahead, separate meetings with all operators were held with the support of Forcam personnel to review the system and its benefits and capabilities. During implementation, meetings were held with small groups of operators to review the Shop Floor Terminal interface and what was expected of the operators.

As much as possible, decisions were made jointly through these informal committees. For example, unlike most shops with monitoring systems in place, Richards Industries does not have large screens in the production area displaying a dashboard or shop report. The consensus among machine operators was that making reports and dashboards only available on desktop computers or laptops at workstations was more useful and less intrusive.

The emphasis on communication is equally important after a monitoring system is gathering data and rendering reports. Mr. Linville makes a point of using the production data as a basis for his daily discussions with supervisors and machine operators. "I can scan a report such as the Operating State Trends report and spot issues such as long setup times or undefined machine stoppage," he explains. "I'll ask about why this or that is happening, but also ask what we can do together to resolve issues and make improvements. We need only five or 10 minutes for this."

Likewise, weekly "Forcam improvement" meetings are now centered on the reports from the monitoring system. As Mr. Linville points out, "When everyone is starting with the facts, discussions can quickly focus on problem-solving suggestions, agreed action items, maintenance priories and so on."What Richards Industries' experience says about the value of communication can be summed up this way: A shop that has good lines of communication in place must keep them open and active when introducing shopfloor monitoring. There should be no surprises. Implementers need to listen and pay attention to intended users. Likewise, a shop can and should find ways to use the data from a monitoring system to strengthen shopfloor communications.

"Monitoring should always be a cohesive force, not a divisive one. That all hinges on good communication," Mr. Metz concludes.

Choose a Champion

There is no question that implementing a machine-monitoring system is a team effort on every level, yet Richards Industries points to Bob Luthy, the continuous improvement/ safety manager, as the main champion of the shop's machine-monitoring initiative. As the leader of the company's ongoing Continuous Improvement Program, Mr. Luthy was in a natural position to take on this role when the pilot program was launched last year.
He served as the key contact person and liaison with Forcam. He is the "go-to guy" for both the shop and the system supplier. He coordinated the effort to match Richards Industries' machine-monitoring objectives with the data collection and reporting capabilities of the Forcam system.

"Mainly, this involved determining what data we wanted on our side and how Forcam could best organize this data and generate the reports that gave us the insights we needed," Mr. Luthy explains.

He coordinated many of the details of installing the system. For example, he and Mr. Linville scheduled when individual machines could be pulled out of production to hook up the network connections with the least disruption.

Once trained thoroughly by Forcam, Mr. Luthy became chief trainer of system users in the shop. For operators, the training covered familiarity with Forcam's Shop Floor Terminal interface at each machine. For supervisors and managers, it covered familiarity with Forcam's Force interface—the suite of reports and performance summaries. As part of this function, Mr. Luthy also acts as chief listener, the person who gathers suggestions

and questions (especially from operators) and presents them to the system supplier for a response. He has also helped the management team understand some of the ways to interpret what the monitoring system was showing about shop performance.

Of course, Mr. Luthy usually leads the review of shop data at the weekly continuous improvement meetings. For example, by going through the previous week's results, the top 10 causes of nonproductive time can be identified and targeted for improvement during the current week. "This is our hit list," he explains.

Mr. Linville credits Mr. Luthy with helping the most to change the shop culture at Richards Industries in a smooth and positive way. "We didn't want machine monitoring to be resisted as spyware, but accepted as a tool that helps everybody," he says. "Bob [Luthy] kept the move to machine monitoring open and transparent. He led the teamwork that made machine monitoring not only acceptable, but embraceable— for operators as well as leadmen, supervisors and managers at all levels."

Good Data Used Wisely

Collecting and analyzing data, sharing reports and displaying results are all pointless if they do not lead to improvements and positive actions. Here are some representative examples of how Richards Industries has benefited from machine monitoring.

Setup reduction. After a few weeks of monitoring, it became apparent that the Okuma LU15 lathe was requiring around eight hours a week for setup, which the operator had tagged as "jaw work." A quick discussion of this issue with the operator revealed that he was routinely reboring hardened chuck jaws to match the diameter of barstock. By acquiring a set of master chuck jaws with soft top jaws in sizes matched to the barstock running on this machine, jaw work has dropped to less than four hours a week.
Less material handling. Encouraged by these results, the shop started looking at the time spent on changing barstock sizes across all of the turning machines. This led to greater attention to grouping orders on the shop schedule by barstock size to reduce change-overs. It takes a little longer to juggle the shop schedule to keep delivery dates on track, but overall time for barstock changes has dropped by about 10 percent in the last three months, Mr. Linville estimates.

"We've made [reducing time for] material handling a priority," Mr. Linville says. "Without machine monitoring to give us a picture of its impact on all 10 machines, we probably wouldn't have targeted this factor for improvement." The goal is to cut the 22 hours of material handling per week by 25 percent in the next three months.

More efficient CNC programs. Barry Haenning, the manufacturing programming manager, took a look at the high number of reported stoppages attributed to "program interrupted." "We found that machine operators were running programs in which the M00 codes for optional stops were still in place," he reports. These optional stops were originally intended to give operators a chance to do spot checks on tool wear or part size when the program was new or had been edited. Once a program had been proven out, these optional stops were unnecessary, but they were rarely removed because hitting the restart button had become part of the operators' reflexes, it seemed. A program to review active CNC programs to delete the stops is underway.

Once a program had been proven out, these optional stops were unnecessary, but they were rarely removed because hitting the restart button had become part of the operators' reflexes, it seemed. A program to review active CNC programs to delete the stops is underway. Recovering overlooked efficiency. One of the most productive machines in the shop is an Okuma MA500 HMC with a Fastems pallet-changing system. Installed in 2007, this cell was to be the prototype for palletization of other new or retrofit machines. Spindle cut time for this cell, however, was only marginally higher than that of non-palletized machines. What was the problem?

"We discovered that we were sending many of our hot jobs to this machine, and we had gotten out of the habit of setting up this work on a pallet offline. We were expediting the job, but not optimizing use of the pallet changer," Mr. Linville says. "Now we are staging work for this machine more carefully so the operator has time to keep pallets loaded and queued for production, including those at the end of the shift for lights-out operation." Spindle cut time is up by 20 percent as a result and steadily moving toward the goal of cutting metal 80 percent of the time during the HMC's 20 hours of daily operation.

Looking Ahead

According to Mr. Metz, Richards Industries will have completed Phase 1 (all machines networked to the monitoring system) by the middle of 2016. He expects Phase 2 (order-data management) to be in place by the end of the third quarter and Phase 3 (paperless shop communication) by the end of the year.

Looking back, he says that the biggest surprise has been finding that the "real numbers" on shop productivity and machine utilization were not as high as managers thought. "We figured that our machines were cutting metal about 50 percent or so over two shifts. It was more like 40 percent when monitoring first started," he reports. It is now higher than 55 percent and climbing.

However, looking ahead, getting the three phases completed will be a plus when business picks up. Mr. Metz and Mr. Linville see three reasons for this expectation. One, machine monitoring is helping the shop unlock capacity that has been hidden by activities other than metal cutting. "We can use data to become a leaner operation," Mr. Metz says.

Two, ramping up production to meet a sales surge will be smoother and more effective. "We can make smarter decisions about allocating shop resources, adding overtime, bringing on new hires or buying equipment," Mr. Linville says.

Three, keeping the workforce stable and engaged will be easier. Mr. Metz believes that moving to a more thoroughly digital environment, in which decisions are closely linked to data, means less stress and smoother interactions on the shop floor. This will also help the company attract and prepare a younger generation of machine operators and programmers. Mr. Linville points out that a significant number of current employees are approaching retirement age.

"We may not be able to replace their skills and experience, but we can create a shop environment where knowledge sharing and collaboration can thrive," he says.

Change and growth have long been key components of the culture and tradition at Richards Industries. This will continue. Machine monitoring is one more step in this direction, and its benefits are likely to create new reasons for its employees to vote the company one of the best places to work in the city.

Chapter 2
Key Performance Indicator (KPI)
Overall Equipment Effectiveness (OEE)

Chapter 2 describes how Key Performance Indicators (KPIs) are used to help a company identify where they need to improve. Those KPIs are used to identify overall inefficiencies, pain points, or specific problem areas that keep them from reaching their goals as a company. The company's focus should be on addressing how to deal with those functionally specific KPI's.

SHOP FLOOR KPI 'S AND OEE

Depending on the company, the focused KPI's will be different per company, industry and mission. KPI's could be as simple as improving the process or as elaborate as those listed below.

Major Shop Floor Issues revolve around:
- Reacting quickly to changing customer demands,
- Improving continuously the quality of products,
- Improving continuously the delivery of products, and
- Reducing costs.

 KPI's could be quality indicators, logistic indicators, and or process oriented indicators.

A company will need to identify how it:

- will visualize processes,
- who will use them,
- who will be the owner,
- what will be the status,
- what will be the goal, and
- what will be the anticipated expectation.

- Implementation of OEE is workable, only if the corporation has committed to go for (CI), "continuous improvement".

The Formula Used for OEE include:

- **Availability (Utilization)**

  Availability takes into account Down Time Loss, which includes any Events that stop planned production for an appreciable length of time (usually several minutes – long enough to log as a trackable Event). Examples include equipment failures, material shortages, and changeover time. Changeover time is included in OEE analysis, since it is a form of downtime. While it may not be possible to eliminate changeover time, in most cases, it can be reduced. By example, utilizing (SME) single minute exchange of dies, technique can reduce the change over time required. The remaining available time is called Operating Time.

- **Performance (Efficiency)**

  Performance takes into account Speed Loss, which includes any factors that cause the process to operate at less than the maximum possible speed, when running. Examples include machine wear, substandard materials, misfeeds, and operator inefficiency. The remaining available time is called Net Operating Time.

- **Quality**

  Quality takes into account Quality Loss, which accounts for produced pieces that do not meet quality standards, including pieces that require rework. The remaining time is called Fully Productive Time. Our goal is to maximize Fully Productive Time! Quality is not an argument anymore; it is a given.

OEE at a macro-level in a plant provides a high level picture of how equipment is functioning which can multiply across linked machines in a series resulting in a picture of the percentage of time expected to get a good part out. This can be used as a signal detector to know where to focus improvement activity at the plant level.

For many companies, Quality is a given at around at least 95% which is standard measurement in industry today. Therefore, the main improvements in terms of OEE can be achieved by increased availability and performance. While availability is purely based on objectively collected machine data, it must be assured that performance is not skewed by wrong assumptions for the planned part time. If a performance of more than 100% is the normal case, and not represented as an outlier, the planned part time in the ERP system has to be adjusted to make the OEE calculation a valuable source of information. All standards that are established must be accurate and achievable.

Chapter 3
Requirements for a True OEE Metric

PLANNED VERSUS ACTUAL PERFORMANCE IN REAL TIME

The calculation of performance in real-time is based on two factors: through a direct connection to the machine the current state of the asset can be monitored and it is always known when a machine actually is producing. Secondly, through the connection to the Enterprise Resource Planning-ERP, system, an operator can choose what order and operation he is currently running on the machine to capture the productive time as well as downtimes while they are occurring. Via manual or automatic input of the produced pieces, the performance for each machine and every order and operation can be calculated via the MES on the fly. This mean that the reporting of down times will be captured through the CNC Controller, PLC and the ERP system.

- **Shop Floor Integration (Assets / Machines)**
  An ERP-Shop Floor Integration framework provides integration of OEE related data with shop floor systems. This framework allows replication of configuration data, master data, and transaction data. Companies have the ability to configure different workflows to process the replicated data.

- **Connecting to Legacy Machines**
  Via a small additional piece of hardware (an I/O Box) it is typically possible to connect to almost any kind of legacy machine. Based on the existing I/Os of a machine, the information can be read and transferred via Ethernet to a data collection point where the information is stored and processed.

That way, even though legacy machines just provide a limited amount of information, in general it is possible to determine the productive state of the machine which is the most important piece of information to gather. A "Productive State" means when a machine is producing product. Often additional states of the machine can be gathered like a program stop or interruption, an alarm or emergency stop or maybe a feed hold.

- **Connecting to New / Advanced Machines**
Newer assets on the shop floor typically come with more advanced software to capture information from the machine or even if it is not a standard function it can be added easily with a software update or upgrade. That way, a wide range of information can be gathered. This contains signals like spindle speed and feed rate, program interruptions, different machine modes like automatic or manual, if an emergency stop has been triggered, or if a cycle was started. Therefore, many different states from such a machine can be determined depending on the combination of signals received.

- **Top Floor Integration (Enterprise Resource Planning)**
Enterprise Resource Planning (ERP) on the shop floor is as critical to the viability of the enterprise as any component in the execution of manufacturing. High-performing manufacturers have found that full integration of real-time operational data collected from the shop floor to the top floor is key to reducing costs, streamlining operations and improving customer satisfaction. Integrated ERP on the shop floor brings manufacturers improved quality, increased production throughput, less scrap, and many more benefits. Companies should have an IT strategy that proves the advantages of using one of the leading ERP software packages focused on linking shop-floor and top-floor operations. ERP systems shed light on manufacturing processes, enabling improved decision-making. For a better explanation of how ERP can interact with MES, see the Success Story on Ultrone. With the integration of the ERP system into the MES, detailed analysis are possible for every machine, as well as operation. A trend can be derived from the performance of an asset as it is attributed to a specific operation. Pareto Analysis, based on materials, can show potentials for improvements. For example, if a certain machine incurs problems when running a certain type of material, we can recognize that there is a need for improvement.

- **ERP Download / Upload**
Based on the information about orders, operations, standard setup and production time, the ERP download is the main requirement to calculate OEE in real-time with the help of the Manufacturing Execution Systems, MES. Feeding back current data via an upload functionality to the ERP system, makes it possible to collect labor time for personnel per order. Whoever is assigned in the company, can update the standard times for orders with accurate data that occurred on the actual machines and with the same operators new orders will run on, making planning much more accurate and reliable.

- **Optimizing Master Data in ERP**
ERP systems can effectively automate a process which greatly improves functionality. ERP systems are not the answer to all of the issues on the shop floor. SFM tools step in where the ERP system are lacking. As a result of using an SFM, this frees an operator to focus on other value-added activities. When additional data is captured, the shop floor is integrated with the top floor for a real-time snapshot of traceability and machine output.

- **Dynamic Finite-Capacity Scheduling**

  Basically, a planning board shows the demand by Work Order. Each line is a work order in a work center. When the machines are loaded, all of the work orders show. But each machine has only so much capacity. Which means it is Finitely Loaded, and visually seen, where bottlenecks are broken down by utilizing other assets that can handle these demand-loads. This action will increase the OEE and the throughput will be increased.

  Through a dynamic scheduling tool that knows all the real-time shop floor data, delays created by breakdowns can be taken into account allowing for an updated plan. Maintenance downtimes can be considered in addition to alternative machines when a bottleneck exists. Having such an automated tool decreases the need for manual interference and provides reliable data.

Chapter 4
The Transparent Factory

Transparent Factories utilize a modular web-based solution for manufacturing order data management, machine data collection, real-time visualization, reporting, detailed scheduling and control. Today, there is no reason to be unaware of what is transpiring on the plant floor, regardless of where you are. This eliminates the hidden factory; those things that are always hidden, not on the operations sheets, or accounted for, in any written procedure. The reason for the transparent factory is to eliminate the hidden factory where the employees' management is watching. This is not about the employee or operator; this is about the process! With an MES, and by understanding what the issues are within the process, the team can resolve the issues much more readily and throughput is increased dramatically. This eliminates non-value-added processes and tasks.

SHOP FLOOR VISUALIZATION AND ALERTING

A very important part of a MES is not only gathering the shop floor relevant data, but making it visible to the right people in the right way. Therefore, visualizations, and reports that alert people about critical issues is crucial for the success of the system.

Web-based Visualization

Visualizations can either be displayed on large LCD/Plasma or PC screens and also allow viewing the current production status from anywhere in the world via a web browser. This is the State-of-the-art real-time production monitoring and visualization.

Individual machines, plants or shop floor layouts can be graphically visualized in real-time, including an integrated status indicator. The display of measurement readings and that of operational states are defined via a graphic, web based, editor. Those visualizations are an important part of an MES. They make information visible in real-time to operators and empower them to see e.g. machine breakdowns right away and act accordingly.

Shop floor visualization is characterized by the flexibility and varying layouts that can be easily constructed to represent the shop floor layout with easy graphics. Status changes in quantity or deviations from the desired limits can be visualized. Predefined or customer specific reports may be integrated in the visualization. The visualization can be seen on an Internet browser, or ANDON device, in the same program the operator is using to interact with the MES to start/finish orders, reports quantities or provides additional information about his machine's state.

These are examples of real time uses of real time data collection, see *Figure 4.1* below:

Figure 4.1 Shop Visualization

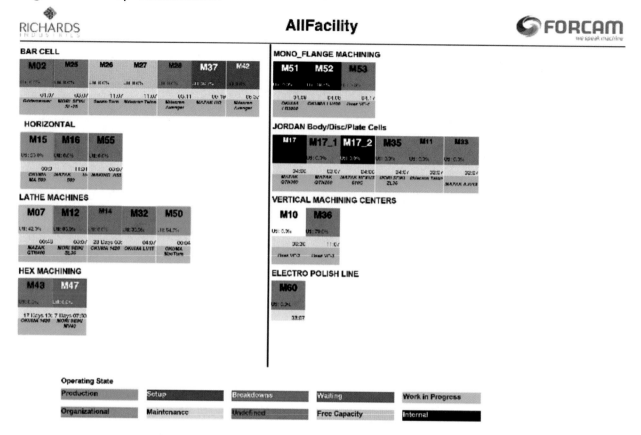

ANDON Displays

Andon is a manufacturing term referring to a system to notify management, maintenance, and other workers of a quality or process problem. The centerpiece is a signboard incorporating signal lights to indicate which workstation has the problem. The alert can be activated manually by a worker using a pull-cord, or -button, or may be activated automatically by the production equipment itself. The system may include a means to stop production so the issue can be corrected. Some modern alert systems incorporate audio alarms, text, or other displays.

An Andon System is one of the principal elements of the Jidoka quality-control method pioneered by Toyota as part of the Toyota Production System, (TPS) and therefore now part of the Lean approach. This gives the worker the ability, and the empowerment, to stop production when a defect is found, and immediately call for assistance. Common reasons for manual activation of the Andon are part shortage, defect created or found, tool malfunction, or the existence of a safety problem. Work is stopped until a solution has been found. The alerts may be logged into a database via a Ticketing system available to the operator as well as supervisors and managers so that they can be studied as part of a continuous-improvement program-(CIP).

The system typically indicates where the alert was generated, and may also provide a description of the trouble. Modern Andon systems can include text, graphics, or audio elements. Audio alerts may be done with coded tones, music with different tunes corresponding to the various alerts, or pre-recorded verbal messages. Nowadays where everybody always has a mobile device nearby, e-mail notifications about production problems took over what used to be transferred via pagers in the past. The use of the word ANDON, originated within Japanese manufacturing companies. In English, this is a loaner word, from a Japanese word, meaning a paper lantern.

MES not only have the capability to send out emails automatically to alert people about malfunctions/ problems on the shop floor, but also an escalation hierarchy can be defined. If for example, after 30 minutes of a breakdown of a machine the supervisor is alerted, it can be said that the maintenance manager is informed after 45 minutes and then the plant manager is alerted after 60 minutes. That way, with increasing loss of productive time, more and more people will be made aware of the   problems in the production run.

ROLE-BASED SHOP FLOOR REPORTING

As stated before, not only the visualization of the gathered data is important, but also that different people have the chance to see data in a different way. For a supervisor, it required to see a shift based report covering every minute of the shift to give him a detailed picture what was going on during the previous shift to prepare for his next upcoming shift. A manager on the other hand is more interested to see a development, maybe of a specific line or the complete shop floor. He should have the option to see reports based on compressed data to get a good overall impression.

Hierarchic Reporting (Machine up to Plant)

Robust production tracking provides tracking of machine production, including detailed and summarized reports. Data is more valuable when it is highly integrated with inventory, tool tracking, and scheduling.

A work-center log provides a detailed history of all events that occur at a machine, including all production, maintenance, and downtime. This function provides reports on uptime, availability, and machine efficiency. Far-reaching benefits result when integrating this function with other business processes including production tracking, inventory, labor tracking, tool-life tracking, and control plans.

Figure 4.2 Machines, Groups and Plants in Hierarchy

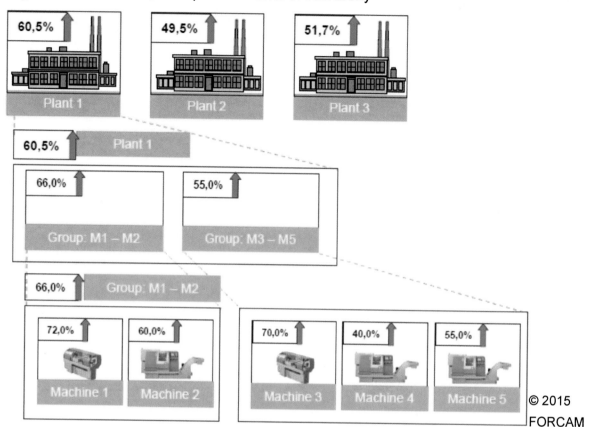

© 2015
FORCAM

A hierarchy ranging from a single machine to covering a whole plant is a prerequisite for role-based reports. A supervisor only wants to see the current data for the machines he is responsible for, while managers are rather interested to get an overall picture and, if possible, even compare different plants in the same division against each other.

The Right Info to the Right Person at the Right Time

MES allows for improved managing and monitoring of information like work-in-process (WIP) on the shop floor, labor and production reporting with the right information in the right time frame. This would include: work orders, workflow management, receipt of goods, shipping, quality control, resource planning, maintenance, finite scheduling, inventory and other areas.

Operator

For the operator, it is important to see the current information of the order he is working on at the moment. He needs to know if he is still on time, compared to the standard time, received from the ERP system. Getting information like, how many pieces he has to produce for a specific order, what material to use and what NC program should be used. A drawing or video might be added to ease his workflow and provide all the information needed to assure on-time delivery of the part.

Product Manager

The product manager might be interested to see how the performance for the latest orders was, if material issues occurred or to understand the reasons for scrap parts. Order, material and quality specific reports created from the data captured via the MES should provide him with this information.

Maintenance

A Pareto analysis about the machines the maintenance team is responsible for with all breakdown states (and reasons) can be very beneficial to determine maintenance intensive machines, specify planned maintenance downtimes and reduce the need for unplanned maintenance stoppages. Based on the data coming from the MES, maintenance will know which parts on a machine need to be replaced on a cyclical basis. By having these parts in inventory, the maintenance department can repair the machine(s) at that time and can reduce the number of days the machine will not be in production.

Based on the data coming from the MES, maintenance will know which parts on a machine need to be replaced on a cyclical basis. By having these parts in inventory, the maintenance department can repair the machine(s) at that time and can reduce the number of days the machine will not be in production.

Industrial Engineering

For an industrial engineer, an MES, is a dream come true. The engineer has a data source through this piece of software giving the individual the option to see all the areas for optimization that might be there at the organization's shop floor. With the desire to improve processes, the engineer can look at the performance of the OEE calculation and the data to improve the NC Program. This can be achieved by digging deeper into the data due to high frequencies of reduced production and non-optimal programs or program stops.

Plant Manager

For a plant manager, the broad picture captured in OEE as well as OEE development is important. The three different parts of OEE are Availability, Performance and Quality. Therefore, the OEE development of the last week and maybe also in comparison to the previous four weeks should be available in his email inbox first thing on a Monday morning. Seeing these trends will be important in developing CIP initiatives.

After deploying the MES, education and training, the team involved should take the time to find out what are the most valuable reports and visualizations for each person depending on their role and focus. This gives them exactly the information they need, potentially automatically sent to them via e-mail to be available exactly when they need it.

For example, before a weekly team meeting, the latest data should be visualized to be discussed or specific reports should be created when focusing on certain target machines or processes.

Figure 4.3 Reporting Dashboards and OEE KPI's

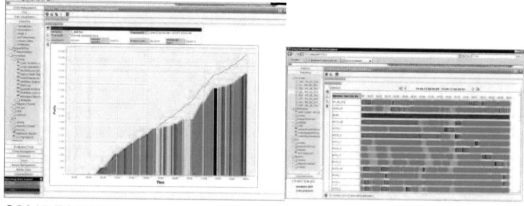

©2015 FORCAM

Figure 4.3 shows on the left an OEE comparison between two different assets. The three different components Availability, Performance and Quality are displayed separately as well as the overall OEE value for both machines and the average value in the right part of this report. This report is helpful for a manager when comparing different machines and evaluating if the goal in terms of OEE can be achieved.

The report in the middle of the figure above, is a Status Gantt Diagram, which shows the states occurring at each machine in a different color. The productivity, displayed in green, is dominating this view, i.e. the availability for the chosen time frame is high. This report is very detailed and can be a great source of information for a supervisor of those machines for example.

Figure 4.4 OEE Dashboards

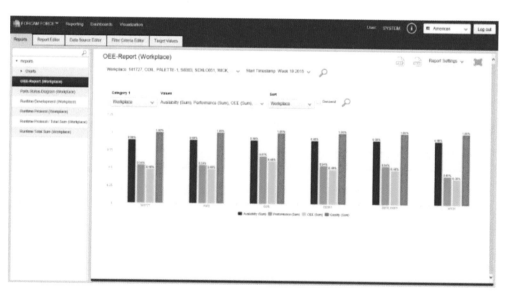

© 2015
FORCAM

On the right the occurred states of one specific asset are displayed over time, while on the y-axes of the graph the produced parts are shown. The black line is the planned part quantity which would need to be met to achieve a performance of 100%. It shows clearly, room for improvement, in terms of performance. For an operator, it is helpful to have the report visible next to his machine. In this way, he is always informed, and able to decide if he can meet the planned part quantity.

Additionally, a combination of reports can be created to display up to date data on screens on the shop floor visible for everyone. The figure below gives, an example of how a combination of reports should appear. Various different data represented is from combined tables and reports which provide an overview of a wide range of information. The displayed example contains everything from short term performance data to the weekly OEE development.

CUSTOMER Success Story—GKN Aerospace
by FORCAM

Engine Systems

This succcess story illustrates how GKN Aerospace Newington has successfully been improving their manufacturing productivity using FORCAM's Advanced Shop Floor Management Technology, FORCAM FORCETM. The following describes how GKN Aerospace management, stakeholders and operators are able to see production anywhere, anytime in real time. The company is now able to analyze machine data and understands current machine states. This allows for timly intervention and quality improvements.

A world of demanding challenges. How does an industry leading company like GKN Aerospace improve their manufacturing productivity?

GKN Aerospace, an aerospace operation of GKN plc based in the UK, is one of the world's largest independent first tier aerospace suppliers, providing complex, high performance, high-value integrated metallic and composite assemblies for aero structures and engine products with a broad range of capabilities. GKN Aerospace Engine Systems, one of four divisions, accounted for 50% of sales in 2013.

GKN Aerospace has close strategic partnerships with all the major OEMs and tier one suppliers – General Electric, Pratt & Whitney Rolls-Royce, Snecma and MTU. This Success Story features the Engine Systems division, a Low- Volume High-Mix/Value Environment of GKN Aerospace, based in Newington Connecticut, USA.

Technology Investment

GKN Aerospace has a reputation for investing in technological innovation, developing and applying new products that bring real and practical benefit to their customers, aircraft operators and passengers. GKN Aerospace's capabilities have been, and continue to be, developed through constant investment and innovation. Their investment in the FORCAM's Advanced Shop Floor Management Technology allows GKN Aerospace to be lean and advances their capabilities to manufacture and assemble metallic engine casings and fixed structures in highest quality standards.

Product Selection

GKN Aerospace leads the market in the design and production of fan containment and non-containment cases in titanium, aluminium, Nickel alloy and composites.

FORCAM FORCE™ is an extremely flexible and powerful tool for improving manufacturing productivity, making it incredibly easy for GKN Aerospace to achieve unprecedented improvements in production. GKN Aerospace envisions new ways to keep customers happy and intends to further increase focus on their needs.

Risks

President, Martin Thorden, says "This technology is a very good tool for continuous improvement activities; we are actually able to see the effect of the change right away and we can now measure our performance on a daily basis - instantly". "This gives us an overview of all of our equipment and how we are utilizing them". When asked about the risk of acquiring the FORCAM FORCE™ system, Martin Thorden proclaimed:

"Strategically, the biggest risk of NOT having this FORCAM FORCE™ Technology is that we would be out performed by our competitors".—*Martin Thorden, President, GKN Aerospace Newington*

GKN Aerospace has had the FORCAM's technology in place for one year. Martin Thorden explains, "We took a moderate approach to what we should be able to achieve in this first year and I am convinced we have been able to achieve our objectives." "Since this is a journey, we are in phase 1 while working towards hooking up to our ERP SAP in phase 2".

In their visual lean manufacturing environment, GKN Aerospace has the FORCAM FORCE™ Technology providing a visual display that shows status information on the plant floor enabling their operators to signal line status based on color: green for normal operation, yellow when assistance is needed, and red when the line is down. This provides increased throughput efficiency, and real-time communication of plant floor status alongside information on Overall Equipment Efficiency (OEE). It is fast and accurate. GKN Aerospace can instantly respond - react - and improve. With focus on continuous improvement, GKN Aerospace is able to see all production in a clear glimpse knowing in real time when lines require attention.

Visibility

Sergio Moren, Director of Operations, at GKN Aerospace Newington states, "The Technology helps me react as well". I've been trying to instill my management team to keep track, while using the system as they learn. They review Gantt charts and reports to see when the machines status starts and stops. This helps us be more proactive as Martin mentioned". "We see benefits that show that we are doing better against our standard hours, which we can directly relate to using the FORCAM FORCE™. We see it in the bottom line".

During the Pilot, the FORCAM FORCE™ technology has helped us get great results, Martin Thorden states. Sergio Moren agrees, "FORCAM is important because everyone knows what to do - we all now know what we have to do. FORCAM helped to quantify and justify our course of action to get to certain levels.

GKN Aerospace has daily production meetings tracking their parts. Engineers are responsible for these parts and need to talk about improvements via their Key Performance Indicators (KPI). KPIs are metrics that facilitate achieving GKN Aerospace's critical goals. The best manufacturing KPIs are aligned with top-level strategic goals to identify and quantify waste.

According to Sergio Moren some KPI's are:

- on time delivery

- quality notification

- toll cost

- inventory

Ewelina Maselek, the Sales and Configuration Engineer mentions KPI's to reduce lead time from about 50 days to 34 days. GKN Aerospace can now run smaller batch-sizes because the flow of production is improved. In gathering the data, "The newer machines can handle more signals so we get more accurate data. As for the older machines, we have to rely on the operator to signal or tell us". In the beginning; it was tougher to trouble shoot signals if the machine was not operating to its full capability. The question always was is it really running 100 percent? A very painful realization.

For instance one machine, a dual ram mill turn with two controllers, was not working properly and we didn't realize it until the operator told us that one had lost connection. There was no automated notification process in place to let us know us about machine status without interaction.

There is a camaraderie between operators to provide feedback. Martin Thorden states, "Part of our success is that we have not forced the data on our operators or held them accountable, they have adopted it willingly as a tool for improvement".

Having a trained empowered workforce makes a difference. This trained workforce could be the difference between a customer putting their work here or elsewhere, stated Sergio Moren.

Reporting and OEE

For GKN Aerospace, Overall Equipment Effectiveness (OEE) is a metric to monitor and improve the efficiency of manufacturing processes. Measured is the percentage of planned and productive production time. Three crucial performance metrics:

Availability, Productivity, Quality

OEE reports are usually only available to managers as a record of what happened in the past. GKN Aerospace can learn from the past but they cannot change it. Now, they can most definitely change the future by utilizing FORCAM's Shop Floor Management Technology. Displaying OEE or related metrics in real-time enables supervisors to proactively set live targets and equips the production team to immediately see the results of their improvement efforts.

The reduction of waste by using the Shop Floor Management Technology, Ewelina Maselek describes as an example:

"We used to have planned stops in our machining process in several 5-10 minutes intervals. We are now able to condense intervals to only a few, which allows us to increase production without the operator present. The Visualization feature of the machine states enables us to see actual run times. By working with the machine operators, we understood that interruption times needed to be reduced. Through data tracking, Ewelina Maselek, is able to notice when a machine's program is not properly updated and running for instance on an old rpm feed rate which does not reflect actual state and leads to reduced production. It is key to make sense of the data provided to take meaningful action.

Sergio Moren details, "Once we had an issue with reduced production because of a new sharpener tool we used for one of our new aluminium containment cases. We were able to spot production time differences. Through investigation with engineering we determined that a dull tool would bring much more productivity and less time for the machine to manufacture. "We are devising improvement solutions".

Carbon Footprint

Sergio Moren, Director of Operations, is currently looking at the reporting module in the office client and doing a run analysis. He is checking the power consumption and evaluating ways to reduce GKN Aerospace's carbon footprint. In this facility, we noticed our highest usage during the summer with all the air conditioners running. Aluminum parts are sensitive to temperature so a certain temperature needs to be maintained. During peak season we correlate with the power company to help us predict usage to offset compressors to not overload the system.

GKN has implemented a collection of improvement techniques based on the idea that waste in all forms should be eliminated from the manufacturing process.

Continuous Improvement

"Continuous improvement is about ongoing vigilance and multiple layers that address machines, people, process and technology enhancements", states President Martin Thorden. Focusing on reducing the time it takes to change a production process from manufacturing one part - to the next enables smaller lot sizes, less inventory, and a faster response to the marketplace, reducing SMED. Sharing information with FORCAM empowers and motivates employees by aligning their efforts with the company. In summary GKN is able to:

- show line status e.g. running, stopped
- give immediate attention to problems during the manufacturing process
- provide simple and consistent mechanisms for communication
- encourage immediate reaction to quality down time and safety problems
- improve accountability and responsibility
- empower operators to take action

ERP Integration

In Phase 2, Ms. Maselek, conveyed that data is gathered to get the FORCAM's Shop Floor Management Technology to integrate with GKN Aerospace's ERP. This integration will allow GKN Aerospace to be even more exact and efficient.

CHAPTER 5
The Smart FACTORY

A smart factory shop floor management tool effectively complements the Continuous Improvement Process (CIP) within any manufacturing facility by collecting real-time objective information from the shop floor and providing the right information to the right people at the right time. This offers the manufacturing environment the needed transparency, synchronization, and standardization.

Information and Communication Technology: The Keys to the 4th Industrial Revolution

Embedded software systems communicate with each other via web-enabled services, and enable:

- Capture of physical data directly with sensors
- Worldwide use of available data and services
- Data analytics and cost savings
- Networking via digital communication technologies (wireless / wired, local / global)
- Human-Machine Interfaces (touch screen, voice control, gesture control)

History - From CIM to MES to SFM

In the Manufacturing Industry, we have seen many changes within the past 40 years. From Flexible Manufacturing Systems-FMS in the late 1970s, we have seen technology mature through Computer Integrated Manufacturing-CIM. FMS islands were then integrated into a system oriented approach to bring the manufacturing facility together for optimum performance and throughput. Even though this was more efficient, there were many platforms and systems running these manufacturing plants.

Enter Manufacturing Execution System—MES. Here we finally realized a more homogenous system that captured more of the factories inputs and outputs that were shared with the rest of the enterprise. As a result, we began to see the emergence of the Enterprise Resource Planning Systems-ERP. Here we finally had all of the enterprise under one roof. Systems like MRP, MRPII, CIM, FMS, and MES started to have common databases and with the advent of open source files, data could be converted and shared by all who needed the data to manage the enterprise.

The advent of the Personal Computer, ASCII, and SQL allowed the end user to create large relational databases where queries could be performed on demand. Hence, knowing what was happening and what the outcomes are, made decision making timelier and resulted in better outcomes.

In the last 40-plus years, we can look back at the aerospace industry at the consolidation of end-players. In the 1970s, there were at least 9 air frame manufacturers. Through the 1980s and 1990s, we saw mergers happen; e.g. Northrup-Grumman, Lockheed-Martin, Boeing bought McDonnell Douglas, and Boeing Integrated Defense Systems (North American Aviation and Rocketdyne) purchased Rockwell International. Sikorski, Pratt and Whitney became part of United Technologies-UTC. This then led to a huge aerospace industry with less players! For these new merger-acquisitions to compete, enterprise systems needed to be designed and developed to handle the huge amounts of data to service their internal as well as their external customers!

The development of ERP systems attempted to do so. In the late 1990s through the first decade of the 21st century, there were many ERP solution providers. As with any new system or product, the respective industry has a 'shaking-out' where the more competitive, and robust product or system, becomes the Order Winners. As a result, the weaker competitors either get bought out or just go out of business.

As this happened, ERP systems added features to their product lines where they tried to have a "One Stop" shopping venue where one solution would satisfy all the needs of the enterprise. We know that each solution provider has their strengths and weaknesses. So when we look to ERP solutions, the question is, can they support and supply the needs of the end user totally? No one system can be exceptional in all of the needs of the enterprise. There may be a need to add, "Bolt-On", a solution to an ERP solution to make it more whole and value-added. This either compliments or supplements a capability the ERP system doesn't have.

So how does this affect Shop Floor Management? Since the 1990s, Just in Time-JIT and then its cousin Lean Enterprises placed challenges on manufacturing facilities to become Leaner, more productive, and most importantly reduce waste. The customer in this new generation of manufacturing would not pay for their products as they did years ago by adding profit to the cost of goods sold-COGS. Today, it is the cost that the customer is willing to pay minus the COGS.

This then changed the margin rules, and companies could no longer pass on the cost of their waste that was realized in their processes. Bad processes lead to variance. Variances lead to manufacturing plants not meeting the customer's needs and wants.

How can a manufacturing facility utilize their ERP system to improve internal operations and on-time delivery of its quality product?

The advent of SFMs adds the necessary elements to improve shop floor performance. Having SFM real-time data available gives Management and the Operator the highest level of performance visibility on the shop floor. Some of the newer ERP systems have SFMs built into them, but there are many "Bolt-On" solutions as well. To this point, the data is converted into "Actionable" Information. By having this data available on production lines, and/or machines, that "go down" will send immediate signals to those who are responsible. Necessary and immediate actions are taken to bring the line and or machine back online. As an outcome, this reduces downtime and improves throughput!

Lean environments require real time data collection, which then leads us in to an Overall Equipment Effectiveness (OEE) program. The OEE system works in tandem with the ERP system that is in use in a manufacturing facility. By selecting a few Key Performance Indictors (KPIs), the production shop concentrates on those and through the collection of data from the shop floor, improves its overall quality and throughput. This is truly the outcomes expected from a Lean enterprise.

In addition, to throughput improvements and the elimination of waste, employees at the point of where work is performed, take on ownership. They feel empowered, are self-directed, work as team players, and have a sense that they are making a difference each day they perform their duties.

Management needs to take charge and be involved and support the efforts of an OEE system. They should have the basic knowledge of such a system and support the entire organization to ensure that this is not another "Flavor of the Month" endeavor. If they embrace it, they will be the "Change Agents" that are needed to make this a success.

Advances in mobile technology has allowed for increasing data availability. The addition of cell phones, tablets, and other mobile devices affords the shop floor to send and receive date via these devices. Bar coding and RFID tags have also improved the way data is transmitted and received. The use of Programmable Logic Controllers (PLCs) and Computer Numeric Controllers (CNC) allow OEE system to be plugged directly into machine tools. Line operations can send valuable information back to operations management so that they can have real time perspective on the output of the manufacturing facility. Employees on the shop floor do not have to input this data and can now digest the information to understand what the process is outputting.

By having this advantage, employees can stop processes that are out of control, find what the root causes are and find corrective actions to bring the process back online. The ultimate goal is to find "Preventive Actionable Solutions" so these incidences do not reoccur!

The use of Statistical Process Control (SPC) can also be utilized to see what the process is producing. SPC is a very powerful tool that allows the operators to visually see processes that are going out of control. The use of Upper and Lower control levels can determine if there is a natural variation in the process and if the variation is central to the control levels. With the elimination of variation, products produced have a higher level quality and ensure that they will perform as designed. In essence, this eliminates defects which are one of the seven deadly wastes of Lean Manufacturing. SPC can also be performed on mobile devices; there are endless possibilities.

Inventory and quality outputs can be sent back to the ERP system to acknowledge the inventory levels and the outcomes of the processes used to manufactures products. A fully-integrated ERP-OEE system is the best solution to providing timely data and improved throughput.

In addition, Total Preventative Maintenance (TPM) can be realized to ensure that machine centers, production lines, cellular manufacturing cells are well maintained. From lessons learned in an OEE system, maintenance systems and departments can determine what the root causes are and how to prevent them going forward. Being able to predict when a machine will fail can eliminate the surprises and then preventive actions can be taken to ensure the maximum up-time! This all leads to a more robust Lean Manufacturing operation and as a result, will lower the COGS then increase the potential margins on products sold. Prices for products can be held, or discounted.

Industrial Internet and Cyber Physical Systems

With the rise of internet use and flexibility, having access to a variety of information increases its popularity. This is not only limited to IT data, but accrual of data from other sources become more readily available in areas, like manufacturing. Topics like the 'Internet of Things' and machine-to-machine communication emerged, i.e. having assets talk to each other and exchange information with other sources. A next step is what is to be known as 'Cyber-physical system' where based on computational intelligence physical entities are controlled. This leads to having networks of elements communicating and interacting with each other.

Advanced Shop Floor Management for a Smart Factory

Through the development of the intelligence of things, machines on a shop floor can become more intelligent and communicate status to other systems. This is the foundation to establish a Smart Factory that can provide an overview of the current status of each machine. It is important to know when a machine is in production, what's being produced and what kind of tool-and-material are being utilized. This allows for precise planning when orders will be scheduled based on data from previous runs of the same item to calculate accurate planned production times, per part. Analyzing delays is easy due to the amount of data available and helping to find ways for future improvement.

Big Data Technology Challenges (disruptions)

Any Browser

The variety of web browsers available is vast these days and should be possible to access the system in any of them.

Any Geographic Location / Time Zones

Companies do not only want to have a MES working on each of their sites separately, they want to compare sites, and divisions, to each other. Due to such a requirement, the geographic location of an asset should not play a part in accessibility, even if reports should be created over different time zones.

Any Language

With globalized manufacturing, a company will not be dominated by one particular language. While it can be expected on a management level that everybody speaks English, it is more likely not the case for other roles like maintenance crews, supervisors or operators. Options to choose different languages can display the same information within in the MES software.

Figure 5.1 Different Language Display

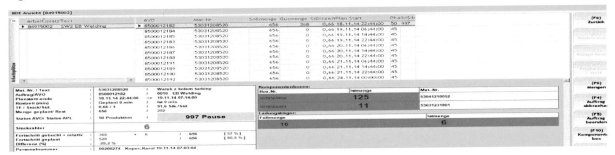

Credit: ©2015 FORCAM

Any Machine

Typically, a new machine is purchased by an organization when needed. The company must take into consideration whether the same manufacturer or controller type exist in their machines. Sometimes, a special machine is needed that is only supplied by a specific manufacturer. This leads to a very heterogeneous set of assets making it impossible to communicate to every machine in the same way when deploying a MES. Therefore, it is a deliverable from the MES provider to connect to any type of machine. With new approaches like MTConnect, a standardized communication for many new machines can be established. However, not all manufacturers support this standard and for almost all legacy machines an update is impossible, or very costly. As a consequence in looking for a MES, one selection criteria should be its ability to connect to the machines present on the organization's shop floor.

Any Device

If you desire to see the real-time data from your shop floor on your computer, laptop, tablet or smartphone, it should always be available, with the same appearance. ME software should be able to support inter-device operability which can be achieved via a web-based technology only using a browser to access and display the data collected.

Customer Story– AUDI
by FORCAM

Overview

Audi is well known as the pioneering car manufacturer in the industry. Audi's press plants in Ingolstadt and Neckarsulm, Germany are equipped with the Advanced Shop Floor Management Technology by FORCAM, which helped increase productivity by 20% during the first year of deployment. In the automotive industry, niche products are becoming more and more important. The automobile industry has changed considerably over the past few years. No doubt: to remain competitive, instant reaction to specific customer requests is of utmost importance. In a parallel to the growing individualization of our society, the range of models available from Audi has increased as well. This, in turn, has implications for the manufacturing of car body panels, as it requires the production of more and more sophisticated pressed parts, involving the use of a wide range of tooling. For Audi, this necessitates creating capacities and improving the efficiency and performance of the press plants.

Car body panels are manufactured for a number of car models at Audi in Ingolstadt and Neckarsulm. More than 1,500 employees work with state-of-the-art press machinery. This independent subdivision is responsible for the distinctive Audi design in aluminium and steel, hence it is one of the most efficient plants in the whole company.

"Thanks to the application software from FORCAM, we are able to completely monitor the production process as well as the necessary infrastructure."
Axel Bienhaus, Head of Central Management / Section: Press Plants, at AUDI AG.

In this respect, the so-called Lean Production has emerged as a resourceful strategy, with constant performance monitoring as an important principle, in order to continuously improve procedures. This ensures constant progress and a sustained high level of productivity. For this, FORCAM GmbH, based in Ravensburg, Germany, is a reliable partner.

Technology

Already during production, all production-related data is available in real-time, allowing for detailed analysis. This facilitates Total Productivity Management (TPM), thus enhancing general productivity and creating important capacities that immediately process production data. FORCAM's solution easily integrates into any ERP with special certification for SAP® allowing companies to link the shop floor - the process level with the top floor, the business planning level. FORCAM's Advanced Shop Floor Management Technology fills the gap between machine control and SAP®ERP.

Pilot

During the pilot phase in 2005, FORCAM implemented the technology in select press lines at the Ingolstadt and Neckarsulm press plant. The modular composition of the solution simplified the gradual implementation of individual modules and allowed for smooth - continuous production.

Rollout

During the same year, FORCAM further rolled out deployment in a total of 33 press lines at Ingolstadt and an additional 15 at Neckarsulm. These press lines are now automatically monitored. Machine signals are captured, data collected and converted into meaningful metrics on any web based device – not just from the shop floor. Shop floor workers can access these visualizations, analyze them and eliminate any deviation from the target. Access to data and information can be customized to each role, which allows for precise address to problems during production in real time. In addition, the solution offers through its standardized data model for comparisons and benchmarking of any plant in any location.

Data is always located on a secured server and available either in the Cloud or Cloud on premises, which offers additional flexibility to Management. The reporting function allows for alerting and even regular reports sent right to the Manager's inbox. Each day, up to 16 gigabyte of raw data is produced, which is handled by standardized compression routines running automatically in the background. Customers can select their Key Performance Indicators (KPIs) such as Overall Equipment Effectiveness (OEE) and receive only information that is meaningful to them.

"The reporting diagrams of the data makes our job a lot easier, because we are now always slightly quicker than before."—*Axel Bienhaus, Head of Central Management / Section: Press Plants, at AUDI AG.*

Harbour Reporting

The Harbour Report is the universally most recognized benchmark for productivity in the automobile industry.

With FORCAM's Advanced Shop Floor Management Technology breakdowns and malfunction are routinely recorded. Any occurrence on machines is registered in the software log and subsequently diagrammed. All data is collected in so-called shift reports, which can be analyzed during the daily meetings of supervisors and shop floor staff.

With the use of Key Performance Indicators (KPIs) and other predetermined metrics detailed reports can be issued that enable domestic and international benchmarking. For Audi in specific select press lines can be segmented in so-called Harbour Categories. Based on acquired data during production, press plant supervisors are able to immediately see how they compare on an international level.

Proven Visibility

Clear and detailed error diagnosis can be conducted based on the setup of the facility. Audi works with large LCD screens, so that each press plant supervisor is able to monitor all machines directly from his office.

Explains Bienhaus, "green means the plant is producing. Yellow means planned maintenance, while blue indicates a new setup. Red or brown suggests malfunction.

Sustainability

Particularly beneficial for Audi is the sustainability of the Shop Floor Management Technology. A seven-level, cascaded, malfunction indicator allows each staff member to clearly identify the cause of malfunction and a defined error message. The first malfunction key indicates whether the breakdown is mechanical, electrical, or tool related.

"For example, if the error is tool related, the subsequent level indicates that the shortfall is due to the upper part of the tool", says Bienhaus. "Eventually, it will become very clear which exact component it was that caused the malfunction."Axel Bienhaus continues, "Malfunction diagnosis gets ever more accurate, right until the deficient element is located."

Productivity

Once Audi had implemented FORCAM's Shop Floor Management Technology clear productivity increases of more than 20 percent were noted.

Audi scores with a far quicker year-end results of all its press plants:

Formerly Audi kept six employees busy over a period of three months, which is now carried out by a single person in just one week.

Section 6
Summary and Conclusion

SUMMARY

This Guide will help you to ask the right questions when considering to purchase a MES (OEE, SFM and ERP). Questions that you and the MES provider should ask should identify if the system purchased is really what matches the Company's needs and expectations. Internal awareness is important to guarantee the success of the system which needs to come with the willingness to have an open mind, to allow changes and strive for improvements.

If you already have a MES, but currently do not see any improvements, the cultural aspect might have been neglected in your organization so far. Try to use this Guide as a guideline to improve the mindset, to be more open to changes and as a way to achieve goals that seemed impossible to reach in the past.

 What you don't know will cause you to fail. You need to answer certain questions.

Some are listed below:

- Is there a demand from the customer?
- Is there a capacity level to consume the demand for a forecast?
- Is there an expected efficiency required to meet the demand?
- Are there issues that prevent you from meeting these goals?
- What are the root causes identified and resolved?
    - Can root-cause analysis be done where the work is performed?
    - Does the data collected by the SFM system allow the company to resolve these issues in an expeditious way?
    - What are the lessons learned in a retrospective view?
    - Is this a genuinely closed-loop system?
    - Does the culture in your organization prohibit change?
    - Is there a focused approach in resolving issues?

CONCLUSION

When thinking about purchasing a MES, (OEE, SFM and ERP) use the same approach as when buying a new asset. Have a precise goal of what could be possible with the tool, based on your requirements and goals. Unfortunately, only having the system deployed does not automatically increase productivity. Therefore, a structured organization with the right mind-set and culture is needed to work with the data on a day-to-day basis. Empowering people is one important step in differentiating roles in the organization and increasing productivity.

An SFM system should be able to connect to all assets in the facility that will be monitored and provide a wide variety of reports supporting different roles in the organization. Additional functions, like the connectivity to the ERP system, and special functions like scheduling, NC program transfer and versioning, personnel time data tracking and/or tracing of parts, can enable a company to get the most benefit from the MES.

What Does this All Mean?

The main thrust of any system is to provide the end user with data that is timely and accurate to make timely decisions. Visibility is of the utmost importance and the availability of data is essential to keep the entire organization engaged so that they are proactive rather than reactive. When issues from the past are repeated, it's too late to think that an organization can recover from such losses. Being proactive, and minimizing such recurrences is key. Using your OEE system with an ERP solution, will create an environment that will have less variation. Inclusion of shop floor personnel will enhance and improve your manufacturing facilities throughput and customer satisfaction. This leads to reduced operating costs and improved bottom line-margins!

Three Factors to Improve an MES

Real-Time Data

Through visualizations the operators have real-time information available that can be a reactive problem-solving approach. That way they can take care of disruptions of the production process right away without delays.

Short Term Data

With the help of the data collected on a daily basis, or per shift, during production meetings, more informed decisions can be made. It can be used to analyze processes, detect room for improvements and check if executed changes lead to the desired results.

Long Term Data

This is the data aggregated over a longer time period, e.g. years. It is important for capital planning, to define bottlenecks and see challenges that occurred over a long time range.

Competition is vicious today and purchasing expensive machines, expecting significant results, without truly analyzing the data of these resources to realize your efficiency, or lack of efficiency, is irrational. Having a competitive advantage in this aggressive environment is driven by managing the true data of these machines and using that information to change. Those companies that truly dentify, have real time correct data on all equipment at all times, and continually improve because of that data, can make the leap into greatness by better utilizing there shop floor resources effectively.

Points Addressed in this OEE-SFM Guide:

- **Offers practical tips for the management of any of today's systems.**

- **Provides an excellent pre-read before any OEE-SFM upgrade or implementation.**

- **Saves money, time and many headaches.**
- **Relevant to the use of an OEE SFM system in a manufacturing plant.**

Good Luck and let us hear from you! We want to know about your successes.

You will get a direct reply from the authors at our website: www.instituteformae.com, Go to *Authors*

Please consider the latest IMAE ERP Guide titled,
"How To Consider, Select and Implement an ERP System."

Step By Step Action Plan GUIDE to Implementation

1. Awareness of pain issues of need to have accurate data available in real time
2. Decision by top management to buy-in
3. Team Creation
   a. Project Manager
   b. IT Specialist
   c. ERP Specialist
4. Preparation
   a. IT infrastructure (server & database)
   b. Machine connectivity & network c. ERP data transfer specification
5. Deployment
6. Go Live
7. Training and start of CIP (work with and analyze data, set a baseline)
8. Start PDCA Cycles

Figure 6.1 Step By Step Plan

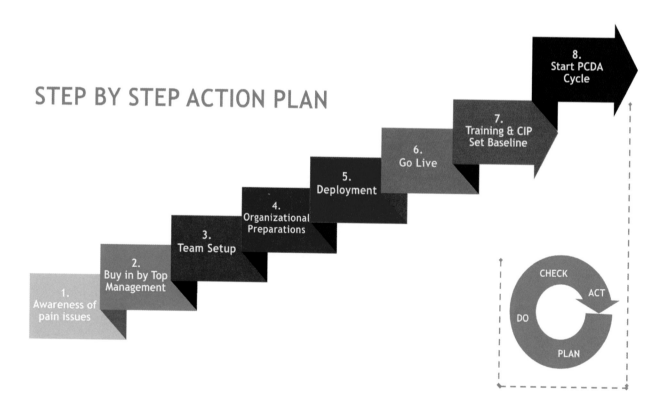

Topics About Which to Think

- ### Involvement of IT

  It is advised to include the IT team early on in the process of deploying a MES. They are the experts when it comes to servers, databases, network infrastructure and security. All of those topics need to be discussed with the MES provider to ensure a smooth and quick installation and configuration.

- ### Involvement of ERP Team

  Just as important as the support from the IT team, is the willingness from the ERP specialists to provide information, time and knowledge. Without a proper integration into the company's ERP system there is no way to automatically calculate OEE with the shop floor management system. The ERP team should be involved from the beginning to have them understand the complete system and its benefits as well as needs when it comes to data transferred from the ERP to the MES.

  If you do not have two separate teams such as an IT and ER P team, you can still reap the benefits of an OEE system on your shop floor by creating a team that is comprised of employees from materials, IT and manufacturing. These individuals can also be subject matter experts (SME's) that can add value to the effort to ensure that the OEE project is a success and meets the expectations of the company stakeholders and ultimate customer.

References / Credits

Articles

- Waurzyniak, P. Managing Factory Floor Data;
- Everett, R.J., Sohal, A.S. (1991) "Individual Involvement and Intervention in Quality Improvement
- Programmes: Using the Andon System", International Journal of Quality & Reliability Management, Vol.
- 8 Iss: 2. Accessed 5 December 2014.

Books

- Liker, J. (2004) "The Toyota Way" New York. McGraw Hill Publishing.

White Papers

- Innovations for Germany, 2014, 53 pages, Order No: 30985

Web References

- www.w3groupllc.com, 2013 http://reliabilityweb.com/index.php/articles/dont_be_misled_by_o.e.e/Dec5,2014 www.plex.com
- http://en.wikipedia.org/wiki/Andon_manufacturing
- http://www.vorne.com
- http://www.oee.com/oee-glossary.html
- http://www.iiconsortium.org/about-us.htm
- http://www.mtconnect.org/
- http://www.oeestandard.com
- Info@Forcam-usa.com

Glossary

- http.//www.oee.com

References / Credits

Credits

Customer Success Stories

FORCAM

FORCAM is a global technology partner to discrete manufacturing companies providing a unified approach for sustained manufacturing excellence across all operations.

FORCAM's awarded technology suite FORCAM FORCE(TM) monitors the performance of over 60,000 machines globally and is deployed by best-in-class companies active in Automotive, Aerospace and Defense, Oil and Gas, and Medical Device discrete manufacturing industries. Customers achieve productivity increases far over 20% in less than 12 months.

FORCAM's Advanced Shop Floor Management Technology harmonizes all manufacturing operations from to the top -- to the shop floor and across the global supply chain and enterprise creating the transparent factory environment.

FORCAM stands for rapid Return-On-Investment (ROI), an increase of Return-On-Capital-Employed (ROCE), and a competitive surge in overall cost savings and reduction of waste. Deliver Results in Productivity.

www.forcam.com

TABLES & FIGURES

OEE GLOSSARY OF TERMS

Good Pieces	Produced pieces that meet quality standards (without rework).	Used in calculating OEE Quality.
Ideal Cycle Time	Theoretical minimum time to produce one piece. The inverse of Ideal Run Rate.	Used in calculating OEE Performance. A variation of the calculation uses Ideal Run Rate instead.
Ideal Run Rate	Theoretical maximum production rate. The inverse of Ideal Cycle Time.	Used in calculating OEE Performance. A variation of the calculation uses Ideal Cycle Time instead.
Lean Manufacturing	Quality philosophy that strives to minimize consumption of resources that add no value to the finished product.	OEE can be a key tool and metric in Lean Manufacturing programs.
Nameplate Capacity	The design capacity of a machine or process.	Used to determine Ideal Cycle Time or Ideal Run Rate.
Net Operating Time	True productive time before product quality losses are subtracted.	Equipment time losses normally are much larger than defect losses.
OEE (Overall Equipment Effectiveness)	Framework for measuring the efficency and effectiveness of a process, by breaking it down into three constituent components (the OEE Factors).	OEE helps you see and measure a problem so you can fix it, and provides a standardized method of benchmarking progress.
OEE Factors	The three constituent elements of OEE (Availability, Performance, and Quality).	Often it is more important to focus on the three OEE Factors than the consolidated OEE metric.
OEE Losses	The three types of productivity loss associated with the three OEE Factors (Down Time Loss, Speed Loss, and Quality Loss).	The goal is to relentlessly work towards eliminating OEE Losses.
Operating Time	Productive time available after Down Time Losses are subtracted.	Operating Time increases as Down Time Losses are reduced.

Performance	One of the three OEE Factors. Takes into account Speed Loss (factors that cause the process to operate at less than the maximum possible speed, when running).	Must be measured in an OEE program, usually by comparing Actual Cycle Time (or Actual Run Rate) to Ideal Cycle Time (or Ideal Run Rate).
Planned Production Time	Total time that equipment is expected to produce.	Benchmark that OEE is measured against.
Planned Shut Down	Deliberate unproductive time.	Excluded from OEE calculations.
Plant OEE	Consolidated OEE calculation as applied to entire plant.	There are different methods of calculating Plant OEE. Pick the one that makes sense for your company.
Plant Operating Time	The time the factory is open and capable of equipment operation.	Planned Shut Down is subtracted from Plant Operating Time to reach the OEE start point - Planned Production Time.
Process	A sequence of activities that starts with some type of input (e.g. raw materials) and ends with some type of output (e.g. a product).	OEE can be used across a wide range of different processes, although it is most often associated with discrete manufacturing.
Production Rejects	Rejects produced during steady-state production. One of the Six Big Losses.	Contributes to OEE Quality Loss (reduces OEE Quality).
Quality	One of the three OEE Factors. Takes into account Quality Loss (parts which do not meet quality requirements).	Must be measured in an OEE program, usually by tracking Reject Pieces.
Quality Loss	Percentage of pieces which do not meet quality requirements.	One of the three OEE Losses (reduces OEE Quality). OEE views defects in terms of lost time.
Reason Code	An identification number or classifiation applied to an Event subcategory. Used to tabulate statistics regarding Events.	Makes it much easier to get a handle on losses, especially Down Time.
Reduced Speed	Cycle where the process is truly running (as opposed to a Small Stop), but is slower than "expected". One of the Six Big Losses.	Contributes to OEE Speed Loss (reduces OEE Performance).

Reduced Speed Threshold	A dividing point between a standard cycle, and one which is considered "slow" (a Reduced Speed cycle).	Setting a Reduced Speed Threshold can be used in Cycle Time Analysis to automatically identify Reduced Speed cycles.
Reject Pieces	Produced pieces that do not meet quality standards.	Used in calculating OEE quality.
Rework Pieces	A subset of Reject Pieces that can be reworked into Good Pieces.	OEE does not make a distinction between pieces that can be reworked and pieces that are scrapped.
Root Cause Analysis	A method of resolving a non-conformance, by tracing back from the end failure to its original (root) cause.	The basic tool for understanding and eliminating the sources of productivity losses.
Run Rate	The production rate when actually producing (running).	Inverse of Cycle Time.
SMED (Single Minute Exchange of Dies)	Program for reducing setup time. Named after the goal of reducing setup times to under ten minutes (representing time with one digit).	Often a part of programs to improve OEE Availability

World Class OEE

- **90% Availability**
- **95% Performance**
- **99.9% Quality**
- **85% OEE**

Made in the USA
Monee, IL
15 October 2020